The Complete ISEE Upper Level Test Prep Book

Over 3000 Practice Questions to Help You Pass Your Exam

CALEB ROSTER

Contents

A GLOBAL PERSPECTIVE.......ERROR! BOOKMARK NOT DEFINED.

A YEAR OF CHANGE.................ERROR! BOOKMARK NOT DEFINED.

POLITICS AND LEADERSHIP....ERROR! BOOKMARK NOT DEFINED.

THE IMPACT OF WW2............ERROR! BOOKMARK NOT DEFINED.

BREAKTHROUGH OF THE YEAR..ERROR! BOOKMARK NOT DEFINED.

PREFACE

What are the benefits of ISEE Upper Level?

Standardized tests are extremely important, and the ISEE Upper Level is one of the most important tests that many private and independent schools employ for admissions purposes. It evaluates a student's intellectual talents and determines whether or not they are prepared to take high school-level classes. If you perform exceptionally well on this test, your chances of being accepted into the institution of your choice will be greatly increased.

The Importance of Being Properly Prepared

In order to achieve success on any standardized test, preparation is essential, and the ISEE Upper Level is not an exception to this rule. Students are tested on a wide variety of topics, including their ability to reason verbally and quantitatively, their comprehension of reading material, their competence in mathematics, and their ability to write essays. The development of the requisite test-taking abilities and confidence can be facilitated by a well-organized and all-encompassing preparation plan, which not only assists in the comprehension of the material being tested.

This Book: A Guide to Navigating It

This book has been organized in such a way that it will walk you through each and every aspect of the ISEE Upper Level exam:

In order to assist you in comprehending the examination, we will first provide an overview of the test, which will include its structure, scoring system, and general methods. The foundation for your preparation is that which is laid out in this part.

The preparation for the ISEE Upper Level is broken down into sections, and each component is discussed in great detail. We will give you with the fundamental ideas, tactics, and a large number of practice questions to put into practice. You will find extensive explanations accompanying each question, which will help you better comprehend the material.

Full-Length Practice Tests: The book contains a significant number of full-length practice tests in order to replicate the experience of taking the actual test. You will be able to evaluate your readiness under situations that are representative of the actual ISEE because these tests are arranged just like the actual exam.

After you have finished the practice tests, you should go back and study the areas that you have already completed in order to reinforce what you have learned. During this section of the book, the emphasis is placed on reiterating important ideas and methods.

Final Preparation: As the day of your test draws near, this section will provide you with last-minute advice, tactics to manage test-day anxiety, and ways to maximize your performance.

Developing a Study Plan for Yourself

Having a well-organized study strategy is necessary in order to achieve success in the ISEE Upper Level. This book will show you how to make one in the following manner:

To begin, you should evaluate your current ability by taking a practice test that is the full length of the actual test. This will assist you in determining the areas that require closer attention.

During structured study, you should set aside a certain amount of time for each portion of the book. Before you attempt the practice questions, you should make sure that you have a solid understanding of the ideas.

Practice and review: It is essential to practice on a consistent basis. Make use of the practice questions to put what you've learned into practice, and go back over the explanations to better comprehend the errors you made.

examinations of Full Length: It is important to take full-length examinations on a regular basis so that you may evaluate your progress and modify your study strategy accordingly.

The next step is to make sure that you have sufficient time to review the material before the day of the test.

By adhering to this organized strategy and making use of the tools that are presented in this book, you will be well on your way to attaining success in the ISEE Upper Level examination. Always keep in mind that the most effective tools you can use on this path are regular work and a happy outlook.

In the following chapters, we will go into each component of the ISEE Upper Level in great detail, providing you with the knowledge, methods, and practice you need to approach the exam with confidence. Your commitment and effort, in conjunction with the advice provided in this book, will be the element that will allow you to realize your full potential and accomplish the academic objectives you have set for yourself.

1

Understanding the ISEE Upper Level Test

The ISEE Upper Level is broken down into its component parts and presented in Section 1.

Students who are interested in entering private high schools, primarily for grades 9-12, are the target audience for the Independent School Entrance Exam (ISEE) Upper Level. This exam is administered to students who are students. The International Student Evaluation (ISEE) is a significant instrument that is implemented by educational institutions in order to evaluate the academic qualities and potential of applicants. Educational Records Bureau (ERB) is the organization in charge of its administration.

The Explanation Behind the Existence of the ISEE Upper Level

The goal of the ISEE Upper Level is to conduct an assessment of a student's ability in a number of different academic fields. The ISEE is a test that evaluates a student's ability to think critically, their aptitude in fundamental areas, and their ability to apply information in innovative situations. When compared to regular school examinations, which test pupils' understanding of particular subject matter, this is a significant departure.

Importance in Regards to the Admissions Process for Schools

There are a number of private and independent universities to which the candidate is applying, and the ISEE score is an essential component of the admissions process for every one of these colleges. It is possible for educational institutions to compare applicants who come from a wide range of educational backgrounds by making use of this uniform set of criteria.

Both the Organization and the Layout of the Second Section

Within the ISEE Upper Level, there are five distinct components that come together to form the whole:

The candidate's vocabulary and their ability to think critically are both evaluated through the use of sentence completions and synonyms in this section of the examination.

The ability of a student to solve mathematical problems without the need for complex calculations is evaluated through the use of quantitative reasoning, which is a sort of analytical reasoning.

During this phase, which is referred to as reading comprehension, readers are evaluated on their ability to comprehend and analyze textual texts.

The candidate's knowledge of mathematical concepts that are regularly encountered in middle school is evaluated in this section, which provides an assessment of the candidate's achievement in mathematics.

Despite the fact that it is not a part of the evaluation process, the essay is distributed to prospective institutions in order to serve as a sample of the student's writing ability.

There is a time limit on each and every section of the examination, and it takes around three hours to complete the whole test.

In the third section, the scores are determined.

Each of the four multiple-choice sections of the ISEE has a score that can vary anywhere from 760 to 940, and the ISEE uses a scaled score method to assign scores. It is not possible to get a penalty for submitting answers that are inaccurate; the points are determined by the number of responses that are correct. The essay is an important component of the application process for the student, despite the fact that it is not scored in any way.

Acquiring Knowledge of the Positions of the Percentiles

Students not only receive scaled scores, but also percentile ranks, which provide an idea of how a student's performance compares to that of other persons who have taken the exam over the course of the previous three years. Scaled scores are the most common type of score that students receive.

Various Methods for Achieving Success on Examinations, Section 4

When it comes to the ISEE, strategies that are successful have the ability to significantly and significantly boost performance. All of the following are included here:

The process of obtaining an understanding of how to effectively allocate time for each segment is what we mean when we talk about time management process.

The process of elimination and educated guessing are two examples of answering tactics. Other examples include techniques such as educated guessing.

In order to be deemed prepared, regular practice with sample questions and full-length examinations need to be considered equally important.

In the area of Stress Management, you will find strategies that will help you regulate your anxiety and maintain your concentration while you are taking the test.

Section 5: Detailed Resources for the Preparation of Activities

Students have access to a wide range of resources, including the ones listed below, which they can utilize in order to effectively prepare for the ISEE Upper Level.

The Official ISEE Guide provides in-depth information regarding the format of the test, in addition to providing examples of questions that may be asked.

It is the intention of the online practice tests to simulate the situations that will be present on the actual test.

Tutoring is a form of specialized teaching that assists in improving areas in which students are deficient.

The term "study groups" refers to conditions that facilitate collaborative learning and offer assistance from other students.

2

Verbal Reasoning

Introduction to Verbal Reasoning is Covered in Section 1

The purpose of the Verbal Reasoning component of the ISEE Upper Level is to evaluate a student's capacity to comprehend and make appropriate use of language. Due to the fact that it forms the foundation for comprehension, communication, and analytical ability, this is an essential attribute for academic achievement.

Structure and The Content

Both of the following components make up the Verbal Reasoning section:

This test evaluates your vocabulary knowledge as well as your comprehension of the meanings of words.

The capacity to employ context cues to finish sentences in a meaningful manner is evaluated through the use of sentence completions.

As a result of the fact that students are needed to respond to forty questions within twenty minutes, time management is an essential ability for this portion.

Key Concepts and Strategies are Discussed in Section 2

It is necessary for students to cultivate a robust vocabulary as well as the ability to recognize context and meaning in order for them to achieve success in the Verbal Reasoning part. These are some strategies:

Building one's vocabulary involves learning new words, their meanings, and how to use them on a regular basis. Instruments such as flashcards and vocabulary lists are helpful in this regard.

The process of practicing sentence completions in order to develop one's ability to infer meaning from context is referred to as contextual understanding.

Learning to recognize and remove replies that are unlikely, with a focus on the choices that are most plausible, is an example of answer strategies.

Practice Questions with Explanations in great detail are included in Section 3.

When it comes to learning the Verbal Reasoning portion, effective practice is absolutely necessary. This section of the chapter has a number of practice questions, each of which is followed by an explanation that is more in-depth. A wide variety of terminology and sentence constructions that may be encountered on the actual examination are reflected in these questions, which have been developed to mirror that range.

Example of a Question for Practice

The word "abandon" most closely refers to the following:

reign of A.

The renunciation of

Assume that C

Conquer the D.

What is the correct answer?

To formally give up or resign a high rank or responsibility is what is meant by the term "abdicate," which is the version of the word that is most closely related to the word "renounce."

Advice for Ongoing Improvement is the topic of Section 4.

Not only does this part contain practice questions, but it also includes suggestions for ongoing progress, such as the following:

When you read widely, you expose yourself to a wide range of materials, such as novels, newspapers, and research articles, in order to become familiar with new vocabulary in its natural setting.

Utilizing Resources: By utilizing mobile applications and online tools for the purpose of expanding one's vocabulary.

Regular review refers to the practice of revisiting and testing oneself on newly acquired terminology on a consistent basis.

Section 5: Final Thoughts and Reflections

The Verbal Reasoning portion requires candidates to have a solid foundation in vocabulary as well as the ability to quickly comprehend language taking into account its context. In order to achieve success in this section of the ISEE Upper Level, it is essential to engage in consistent practice, to make strategic preparations, and to place a strong emphasis on developing a broad vocabulary.

3

Quantitative Reasoning

Within the first section, a general introduction to quantitative reasoning is provided.

The objective of the Quantitative Reasoning section of the ISEE Upper Level is to assess a student's capability for mathematical reasoning as well as their ability to find solutions to problems. Not only will you be asked to calculate solutions in this area, but you will also be expected to comprehend and evaluate mathematical concepts before moving on to the next section.

Organization and the Content of the

The following are the two primary categories that are included in this section, which consists of 37 questions, each of which must be answered within a time limit of 35 minutes:

A person's ability to apply mathematical principles to circumstances that occur in the real world can be evaluated through the use of word problems, which are a form of question.

As part of the quantitative comparisons, students are needed to compare two different quantities and determine the relationship that exists between them. This is a must.

Section 2 offers a comprehensive discussion of fundamental mathematical ideas.

A wide range of mathematical topics, including the following, are studied within the Quantitative Reasoning portion of the curriculum.

Algebraic phraseology

In terms of geometry,

An Examination of Numbers

Probability of events taking place

An Analysis of the Data

In order to successfully complete the tasks that are outlined in this section, it is vitally necessary to have a comprehensive understanding of the concepts that are now being discussed.

The third portion will provide you with practice questions along with thorough solutions to those questions.

For students to achieve a high level of proficiency in the Quantitative Reasoning section, they will need to devote a significant amount of time to practicing. This section has a wide variety of practice questions, ranging from elementary to higher levels, and it is included in this subsection. The answers to each of these questions are followed with an explanation that is both detailed and easy to follow. It is beneficial for students to utilize this method because it enables them to comprehend not just the "how," but also the "why" behind each challenge.

An Illustration of a Question Useful for Practice

Are you able to provide me with the value of x in the event that the equation $5x + 3 = 2x - 4$ is accurate?

Therefore, x is equal to -7/3.

It is recommended that you begin by arranging all of the x terms on one side of the equation, and then proceed to arrange the constant terms on the opposite side of the equation. When 5x is subtracted from 2x, the result is -4 subtracted from 3. Three times equals seven, to simplify the situation. The answer to the equation $x = -7/3$ is obtained by dividing both sides by three.

The topic of discussion for Section 4 is various approaches to problem solving.

Not only should students get experience in finding solutions to problems, but they should also focus on developing efficient methods for finding solutions, such as the ones discussed in the following paragraphs:

Before attempting to locate a remedy to the problem, it is essential to conduct a comprehensive reading of the issue and make certain that you have a complete understanding of what it is asking of you regarding the solution.

There is a procedure known as estimating, which involves the rapid examination and removal of faulty answer options through the utilization of estimation.

An example of working backwards would be to begin with the response alternatives that are provided and then work your way backwards towards the question.

Through the process of acquiring the ability to make a snap decision about whether to solve a problem or to move on to another work and come back to it later and if time permits, time management is the process of developing the capacity to make a decision.

Section 5 is devoted to the topic of advice on continuing one's education.

The following are elements that must be present in order for quantitative thinking to continue to advance:

Consistently finding solutions to a wide variety of problems is what we mean when we talk about "regular practice."

The act of seeking assistance is the action that one takes when they are having difficulty understanding an idea.

In the context of examining mistakes, the term "learning from mistakes" refers to the process of collecting knowledge about how to recover from mistakes.

An activity that is pleasurable to participate in is playing mathematical games and solving mathematical puzzles that test one's ability to think mathematically.

Final Thoughts and Reflections are Presented in Section 6

It is not sufficient to simply possess mathematical abilities in order to achieve success in the Quantitative Reasoning section; rather, it is required to employ these abilities in a manner that is both intelligent and strategic in order to accomplish success. It is crucial to engage in consistent practice in order to attain success in this field. In addition, it is necessary to master effective strategies and have a comprehensive understanding of fundamental ideas through practice.

4

Reading Comprehension

How to Understand Reading Comprehension is Covered in Section 1

The Reading Comprehension component of the ISEE Upper Level is designed to assess a student's capacity to read, comprehend, and evaluate written content. The critical thinking and the capacity to infer meaning from text are both talents that are crucial for academic achievement, and this portion examines both of those abilities.

Structure and The Content

In this portion, you will be given thirty-five minutes to complete the thirty-six questions that are based on the six passages. There are many different subjects that are discussed in the chapters, such as social studies, science, and literature. Main ideas, supporting information, inferences, and terminology in context are all included in the questions that are being tested.

Effective Reading Strategies are discussed in Section 2.

For pupils to achieve mastery in Reading Comprehension, they need cultivate:

By forming predictions, asking questions, and summarizing information, active reading is a kind of reading that involves engagement with the text.

While reading, taking notes involves writing down important topics, themes, and words that are foreign to you.

Having a vocabulary that is contextual means having the ability to understand the meaning of words based on the context in which they are used.

Making inferences is the process of drawing conclusions based on the information that is presented in the text.

Practice passages and questions with clear explanations are included in Section 3.

In order to become proficient in the Reading Comprehension section, practice is essential. Several practice passages are provided in this section of the chapter, and each one is followed by a set of questions that are similar to the types of questions that are encountered on the ISEE. Students are better able to comprehend the reasoning behind the correct answers and how to tackle questions of a similar nature in the future when they are provided with detailed explanations for each topic.

An Example of a Question and Passage Passage Practice

This passage is a tiny excerpt from either a fictitious narrative or an article that is not from fiction.

What is the most important point that the passage is trying to make?

Explanation and Answer: The answer will be determined by the passage, and the explanation will demonstrate how to determine the primary idea by analyzing the various details and themes that are present throughout the passage.

Advice for Ongoing Improvement is the topic of Section 4.

The following are the components of ongoing improvement in reading comprehension:

Reading a wide range of texts on a consistent basis, including works of fiction, non-fiction, and academic articles, is referred to as diverse reading.

Discussing and analyzing what you have read with your instructors, classmates, or family members in order to gain a deeper grasp of the material.

The process of continually acquiring new words and how to use them is known as vocabulary expansion.

Simulation of test settings in order to enhance both speed and efficiency is what is meant by "practice under timed conditions."

Section 5: Final Thoughts and Reflections

During the Reading Comprehension section, you will not simply be required to read rapidly; rather, you will be expected to read with an analytical mindset. A constant practice routine and exposure to a wide range of reading materials are necessary for the development of abilities such as comprehending the text, correctly understanding its meaning, and providing proper responses to questions. Students will be provided with the resources and tactics necessary to achieve success in this crucial area of the ISEE Upper Level following the completion of this chapter.

This chapter section, which is approximately 430 words in length, offers a comprehensive overview of the Reading Comprehension section of the ISEE

Upper Level test. It places an emphasis on the significance of efficient reading techniques, sufficient practice, and a variety of reading habits in order to achieve success in this section of the exam.

The user

proceed to the following chapter.

The ChatGPT

The Mathematics Achievement Section is covered in Chapter 5.

Fundamentals of Mathematics Achievement is the topic of Section 1.

The purpose of the Mathematics Achievement component of the ISEE Upper Level is to evaluate a student's level of skill in mathematical subjects that are normally taught in middle school. This section goes beyond simple problem-solving to assess the student's comprehension of mathematical principles and practical application of those principles.

Structure and The Content

Covering a wide range of topics, this section consists of forty-five questions that must be answered within forty minutes.

Algebraic terms

The geometry of

Numerical Analysis

An Examination of the Data and Probability

The act of measuring

There are a variety of tasks, ranging from simple mathematics to more complicated ones that need analytical thinking.

Techniques for Solving Problems at an Advanced Level, Section 2

For students to be successful in the Mathematics Achievement section, they need to have a strong comprehension of mathematical ideas and the ability to apply those concepts in a variety of different situations. Here are some important methods:

Understanding concepts means having a firm grasp on the fundamental ideas that underlie various mathematical disciplines.

Understanding when and how to apply formulas and theorems is referred to as "application of formulas."

Using logical thinking to find solutions to difficult situations is an example of logical reasoning.

The process of using alternative methods to verify responses is referred to as cross-checking.

In the third section, you will get comprehensive practice questions along with their solutions.

It is essential to practice solving a wide variety of problems in order to achieve mastery of the Mathematics Achievement portion. This section of the chapter has a number of practice questions that cover all of the concepts that have been discussed, and each of these questions is accompanied by a comprehensive solution. With this strategy, students are better able to comprehend the approaches that are used in the resolution of a wide variety of mathematical issues.

Example of a Question for Practice

To find the value of x in the equation $2(x - 3) = 4(x + 2) - 10$, please solve the equation.

The answer and solution are as follows: the solution entails expanding the problem, gathering terms that are similar, and solving for x, which ultimately results in x equaling 5.

Techniques for Review and Reinforcement are Discussed in Section 4

Students ought to do the following in order to promote learning and enhance performance:

Regular review entails going back over fundamental ideas and equations on a regular basis.

Practice exams: The process of simulating the environment of the examination by taking timed practice exams.

Understanding and gaining knowledge from mistakes made in practice issues is what is meant by "mistake analysis."

Collaboration with peers for the purpose of gaining a variety of viewpoints and approaches to problem-solving is the focus of peer study groups.

Section 5: Final Thoughts and Reflections

The Mathematics Achievement portion is not just a test of a student's ability to calculate, but it also evaluates the student's capacity to think mathematically. In order to achieve success in this area, it is necessary to have a solid understanding of the concepts, to successfully apply those concepts, and to practice solving a range of situations. Students will be able to improve these skills and reach a high degree of mathematical competency for the ISEE Upper degree with the help of the guidance and tools that are provided in this chapter.

5

Essay Writing

The First Section: The Fundamentals of Essay Writing

The Essay Writing section of the ISEE Upper Level is an important component that gives students the opportunity to demonstrate their writing skills, inventiveness, and capacity to organize their views in a cohesive manner. The essay is a major part of the application, despite the fact that it is not evaluated like the other portions. This is because colleges use it to evaluate a student's writing abilities and potential for success in an academic environment that is challenging.

Structure and The Content

Students are required to react to a question that is provided in the essay part within a time limit of thirty minutes. According to the prompts, students are often required to compose an essay on a specific subject that is either narrative, expository, or persuasive. This gives them the opportunity to demonstrate their writing style, reasoning, and command of language.

Analyzing Essay Prompts is the Topic of Section 2

It is essential to thoroughly comprehend the essay challenge and to react to it in an appropriate manner. The following are important steps:

The process of analyzing a prompt involves determining the sort of essay that is required (narrative, persuasive, or expository) and comprehending the questions that are being asked by the prompt.

The process of coming up with ideas and organizing thoughts before to beginning to write is known as brainstorming.

The process of developing a thesis statement that is crystal clear and serves as a compass for the essay.

Writing Methods and Organization are Discussed in Section 3

Included in a well-structured essay are the following:

The thesis statement is presented in the introduction, which also establishes the tone for the rest of the essay.

There should be a single point that is discussed in each of the body paragraphs that supports the thesis. Examples and facts should be used to support this point.

In the conclusion, a summary of the most important issues is provided, and the thesis is restated in light of the evidence described.

Other components of effective writing include:

Coherence and Clarity: Writing in a way that is both clear and logical, making sure that each paragraph flows into the next without any interruptions.

Using a wide range of words and sentence structures is an important part of vocabulary and style.

grammatical and mechanics refer to the practice of paying attention to spelling, punctuation, and grammatical use.

Evaluations of Sample Essays are Presented in Section 4

In this part, sample essays that are written in response to typical prompts are provided with the purpose of assisting readers in comprehending what constitutes a good essay. The critique that is sent with each sample focuses on the content, organization, style, and mechanics of the sample, highlighting both the areas in which it excels and those in which it could be improved.

Tips for Effective Essay Writing is the topic of Section 5.

In order to improve one's essay writing skills, one must:

Essay writing on a variety of subjects should be done on a regular basis in order to acquire versatility.

Inquiring about comments from peers, professors, or tutors is an example of feedback.

It is important to read a wide variety of texts in order to gain an understanding of the various writing styles.

The practice of writing essays within the allotted thirty minutes is an example of using time management.

Section 6: Final Thoughts and Reflections

Students have the extraordinary chance to establish a personal connection with the admissions committee through the Essay Writing component of the application. It is not enough to simply demonstrate that you are proficient in writing; you must also exhibit the ability to communicate your views, opinions, and personality. Students have the ability to compose essays that are memorable and have a favorable impact on their ISEE Upper Level grades if they engage in practice, receive feedback, and concentrate on teaching effective writing strategies.

6

Practice Questions

This section will provide an overview of the significance of practice exams in Section 1.

The process of preparing for the ISEE Upper Level includes practice tests, which are an integral component of the preparation process. With the assistance of these, students are able to develop a greater sense of self-assurance, identify areas in which they may improve, and become more acquainted with the structure and timetable of the examination. In this section, you will find a variety of full-length practice examinations that are intended to simulate the experience of taking the actual examination.

The first practice test is covered in Section 2.

The purpose of the first practice test is to serve as a simulation of the actual ISEE Upper Level assessment, both in terms of its organization and the level of difficulty it presents. It includes not just sections on verbal reasoning and numeric reasoning, but also sections on reading comprehension, mathematical performance, and an essay prompt. In addition, it includes sections on analytical thinking and logical reasoning.

The content of the instructions includes specific directions on how to approach each subject in detail.

When it comes to timing, each particular section is timed in a manner that corresponds with the actual conditions of the examination.

For the purpose of providing a record of their responses for future reference, each student is given a one-of-a-kind answer sheet.

The second practice test is included in Section 3.

The second practice test is organized and timed in the same manner as the actual examination, despite the fact that it comprises a different set of questions than the first practice test. The purpose of this is to provide the learner with additional challenges that will aid in their learning and preparation.

A wide range of new questions are used to evaluate the student's ability to adjust to new situations and the breadth and depth of their information acquired.

The Essay question is a one-of-a-kind question that might be utilized to enhance one's writing skills throughout the entirety of the essay portion of the test.

The Third Practice Test is a component of Section 4.

While the third practice exam continues to give a wide range of questions and problems, it also ensures that all of the subjects and question types that are anticipated to be encountered in the ISEE Upper Level are covered in their entirety.

There is a possibility that some of the questions will be slightly more challenging than others in order to evaluate the student's capacity to apply concepts while under duress. These questions are intended for advanced students.

Section 5 contains both the explanations and the answer keys for the questions.

Following the completion of each practice exam, detailed answer keys are provided, along with explanations for each question found on the test. For the purpose of comprehending errors and acquiring the knowledge necessary to address concerns of a similar sort in the future, it is crucial to have these explanations available.

When it comes to problems that are quantitative in nature, detailed answers are provided in a format that is step-by-step.

Explanations for correct answers, including the reasons why alternative choices were incorrect, are included in the insights for both the reading and speaking portions of the test.

Your performance evaluation is covered in Section 6 of this document.

Following the completion of each test, the students should conduct a performance evaluation in order to identify the areas in which they excel and those in which they have room for improvement. There is information provided on how to evaluate the outcomes of tests and how to make use of those results to improve the techniques of study.

The act of recognizing different kinds of problems that are commonly challenging is referred to as "identifying patterns."

When doing a review of time management, it is necessary to determine whether or not the allotted amount of time was sufficient for each component, and then to adjust your methods in accordance with the findings.

Final Thoughts and Reflections are Presented in Section 7

One of the most effective ways to get ready for the ISEE Upper Level is to regularly practice answering questions that are of a complete length. This is one of the most efficient ways to prepare for the exam. Not only does it improve one's knowledge and abilities, but it also helps reduce the amount of worry that one feels on the day of the examination. For the purpose of ensuring that they are prepared for the real thing, students should take these practice exams seriously and make an effort to reproduce the conditions of the actual exam as accurately as possible.

7

Review

Key Takeaways from Each Section are Discussed in Section 1

The most significant ideas and methods discussed in each chapter are summarized in this section of the book, which is included at the end of the book. As a quick-reference guide, it serves the purpose of reinforcing the most important topics that are required for success on the ISEE Upper Level.

This section will review the most important vocabulary and sentence completion skills for verbal reasoning.

An examination of the fundamental mathematical ideas and approaches to problem-solving is included in the scope of quantitative reasoning.

A comprehensive overview of good reading strategies and ways for responding to a variety of queries is provided in the section on reading comprehension.

A consolidation of the most important themes and formulas in mathematics is referred to as achievement.

The following are some key points to consider while writing an essay that is well-structured and cohesive.

Additional practice questions are included in Section 2.

Furthermore, this part offers additional practice questions for each section of the ISEE Upper Level, in addition to the full-length practice tests that are already included. The purpose of these questions is to provide additional practice while also focusing on common weak spots in the curriculum.

A wide variety of questions that cover all of the necessary abilities and subjects are included in the diverse question types.

Explanations in Great Detail: In order to facilitate increased comprehension, each question is accompanied by an explanation.

Tips for Ongoing Education is the topic of Section 3.

Providing guidance on how to continue studying and getting ready for the ISEE Upper Level is the purpose of this part, which is intended to ensure continuous improvement.

The strategies for routinely examining key topics and practice questions are referred to as "regular evaluation."

Information regarding supplemental materials and resources, such as online practice tools and tutoring services, are included under the category of "Additional Resources."

In this section, we will discuss how to stay motivated and how to manage stress while you are on the journey of preparation.

Personalizing Your Study Plan is Covered in Section 4

This section offers instructions on how to tailor the study schedule based on individual performance in practice tests and exercises. This is done in recognition of the fact that every student has characteristics that are unique to them, such as strengths and limitations.

Finding and concentrating on one's own areas of weakness is what is meant by the term "personalized focus areas."

The term "balanced preparation" refers to the process of ensuring in-depth preparation that encompasses all aspects of the examination.

Section 5: Final Thoughts and Reflections

It is essential to conduct a review and reinforcement phase in order to consolidate one's knowledge and skills before to the examination. This chapter urges students to adopt a proactive approach to their preparation by routinely reviewing and practicing essential ideas, as well as continuously adjusting their study strategy, in order to ensure that they achieve the highest possible score on the ISEE Upper Level.

www.ingramcontent.com/pod-product-compliance
Lightning Source LLC
Chambersburg PA
CBHW061634130726
47996CB00003B/1283